MEDIA GUIDE for Stud

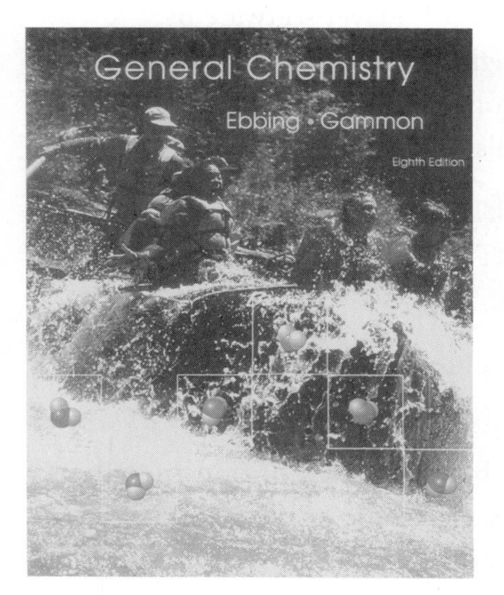

To accompany

General Chemistry, Eighth Edition

Darrell D. Ebbing
Steven D. Gammon

Houghton Mifflin Chemistry
Flexible, integrated learning tools

This handy guide provides information and access to the media resources available with the Eighth Edition. These tools can help you succeed in your course.

Contents

Printed in the U.S.A.

Visit: chemistry.college.hmco.com/students
For Technical Support: (800) 732-3223 or support@hmco.com

Master concepts.

Study more effectively.

Practice problem solving.

The MEDIA GUIDE helps you succeed in your course.

MOLECULAR ANIMATIONS AND CONCEPTS

- **Interactive Student CD-ROM**
- **Student Website**

INDEPENDENT STUDY AND REVIEW

- **Interactive Student CD-ROM**
- **Student Website featuring SMARTHINKING™ live online tutoring and flashcards**

ONLINE HOMEWORK

- **Student Website featuring ACE quizzes**
- **Eduspace®** (powered by Blackboard™)**: featuring online homework with end-of-chapter problems from the text**

Visit: **chemistry.college.hmco.com/students**
For Technical Support: **(800) 732-3223** or **support@hmco.com**

SMARTHINKING™ live online tutoring

Best Uses

- Accessing live tutorial help outside class
- Reviewing difficult concepts

If you don't have time to see your instructor, or office hours aren't convenient, this is your chance to get help when you need it.

This live, online service provides personalized, text-specific tutoring when you need it most. **With *SMARTHINKING*™ you can:**

- Connect immediately to live help during typical study hours: Sunday through Thursday from 2 p.m. to 5 p.m. and 9 p.m. through 1 a.m. EST.
- Submit a question to get a response from a qualified e-structor, usually within 24 hours.
- Use the whiteboard with full scientific notation and graphics.
- Pre-schedule time with an e-structor.
- View past online sessions, questions, or essays in an archive on your personal academic homepage.
- View your tutoring schedule.

E-structors help you with the process of problem-solving rather than supply answers.

▼ Whiteboard

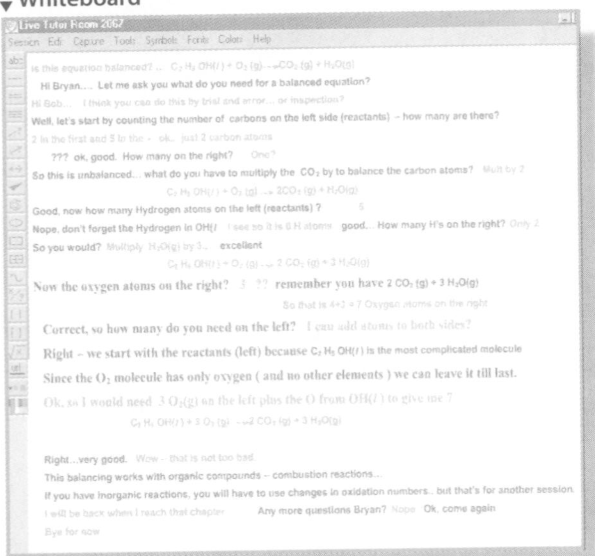

▼ SMARTHINKING™ live online tutoring homepage

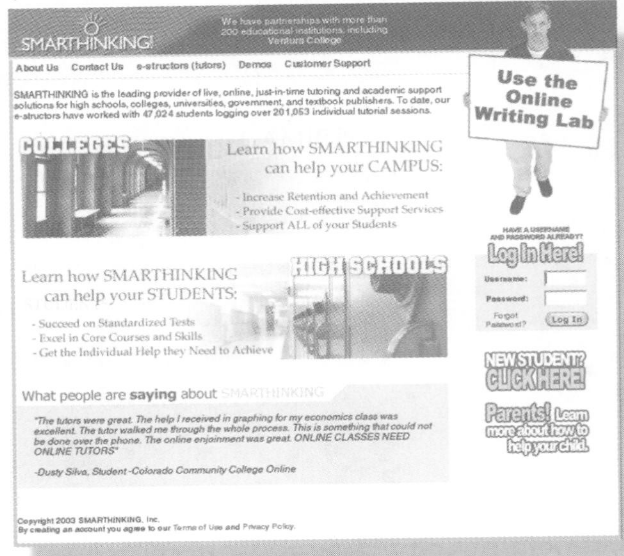

Visit: **smarthinking.com/houghton.html**

For Technical Support: **(888) 430-7429, ext. 1** or **info@smarthinking.com**

Logging On

1. Go to smarthinking.com/houghton.html.
2. Follow the instructions online to set up your student account using the passkey located on the inside back cover of this guide.

For technical support for *SMARTHINKING™*,
call **(888) 430-7429**, extension 1
or
email **info@smarthinking.com**.

If your instructor has adopted *Eduspace®*, you may access *SMARTHINKING™* tutoring through *Eduspace®* by going to www.eduspace.com.

- Once you have registered in *Eduspace®* (by following the instructions on page 11 of this guide), go into your instructor's Ebbing/Gammon, *General Chemistry*, Eighth Edition course.
- Click the Tutoring button on the left side of the screen. You will automatically be taken to *SMARTHINKING™*. You don't need to additionally register or log in to that site.

Student Website

Best Uses

- **Practicing problem-solving**
- **Reviewing chapter topics**
- **Visualizing chemical concepts at the molecular level**
- **Accessing study strategies and text related material**

Organized by chapter, the ***Student Website*** supports the goals of the Eighth Edition with the following resources:

Visualization, Practice, and Tutorial Tools:

- Interactive Understanding Concepts and Visualizations tutorials that use images, animations, and videos to help reinforce core chemical concepts. (Also found on the Student CD-ROM.)
- Flashcards of key terms and concepts.
- Houghton Mifflin's ACE self-quizzing.
- Molecule library.

Additional Resources and Study Aids:

- An interactive periodic table.
- Strategies and solutions for all Concept Checks in the text.
- Glossary from the text.
- Link to *SMARTHINKING*™ online tutoring.

Interactive flashcards

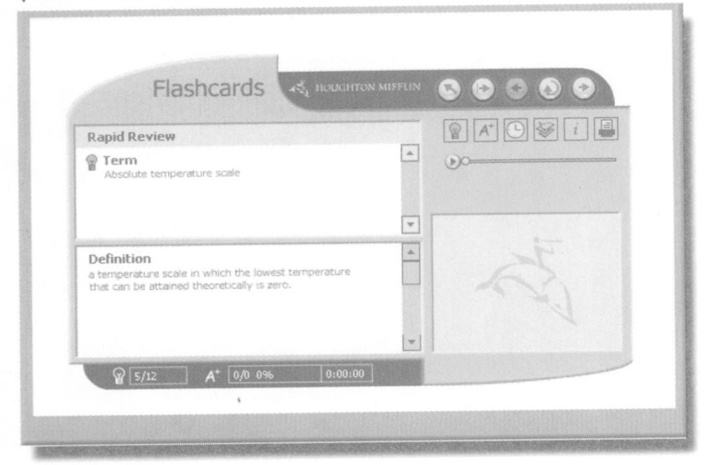

ACE self-quizzes

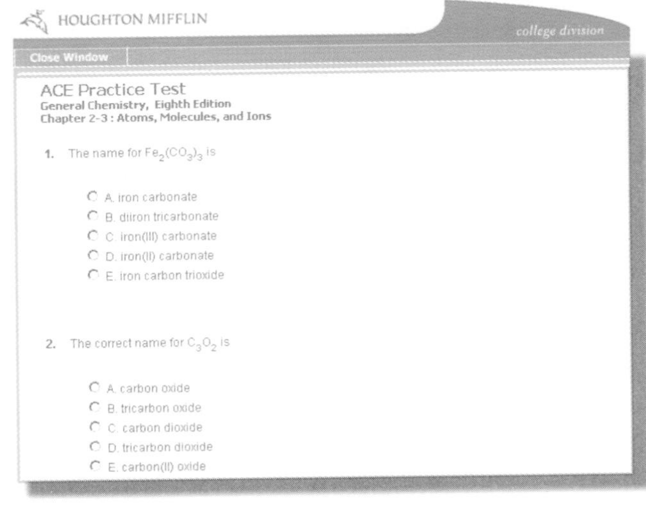

Visit: chemistry.college.hmco.com/students

For Technical Support: (800) 732-3223 **or** support@hmco.com

Student Website

Logging On

1. Go to chemistry.college.hmco.com/students.
2. Select your text.
3. Enter the user name and password located on the inside back cover of this guide.

If your instructor has adopted *Eduspace®*, you may access the Student Website through www.eduspace.com. Once you have registered for *Eduspace®* (by following the instructions on page 11 of this guide) and entered your instructor's Ebbing/Gammon, *General Chemistry*, Eigth Edition course, click Course Materials and then click the Student Website link.

▾ Student website homepage

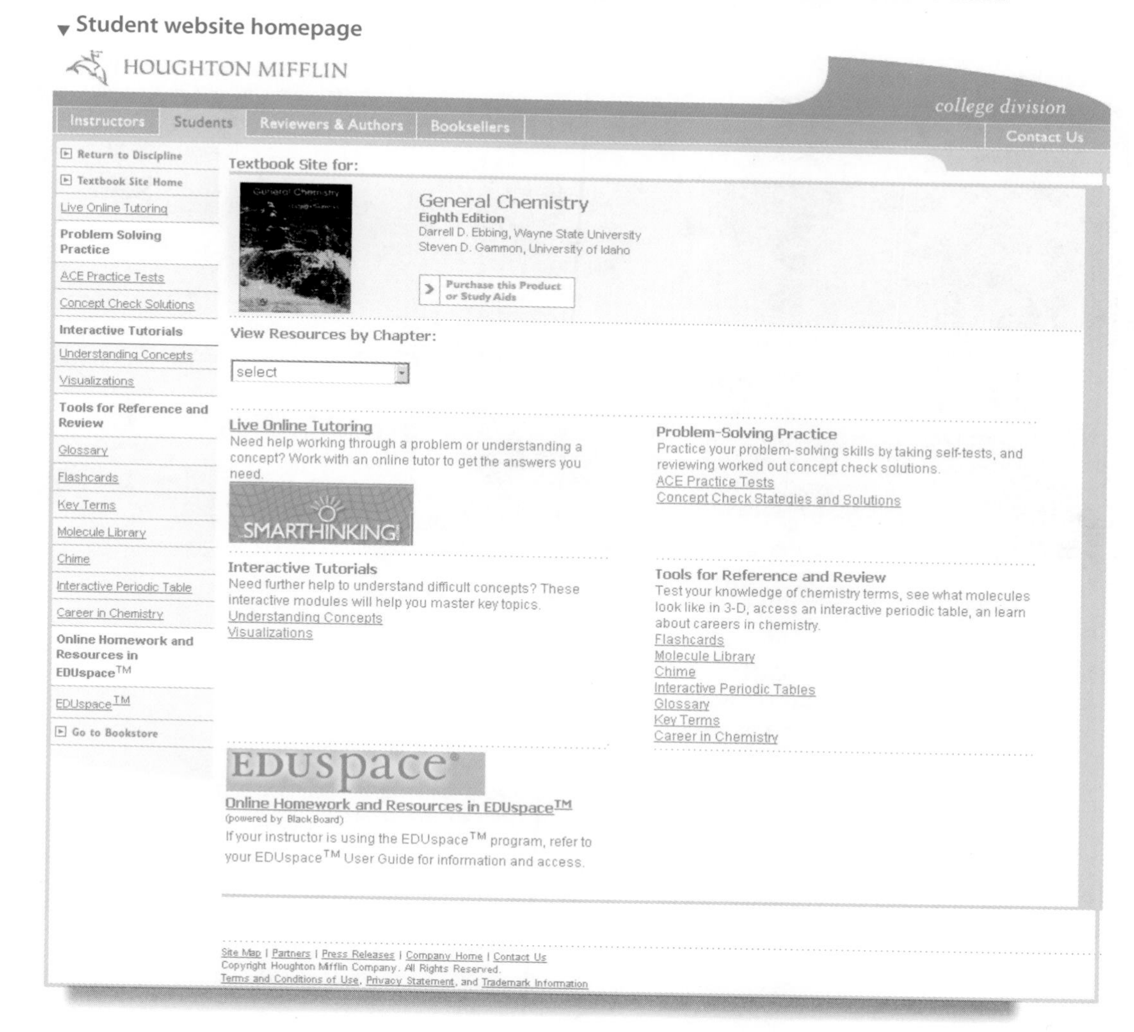

Student CD-ROM

Best Uses

- Reviewing chapter topics
- Visualizing chemical concepts at the molecular level

This comprehensive CD-ROM contains text-specific support and materials to help you study outside of class.

The main menu provides links to these valuable features:

- Interactive Understanding Concepts and Visualizations tutorials that use images, animation, and videos, to help reinforce core chemical concepts.
- Overviews of key topics in each chapter.
- An interactive version of the periodic table.
- Key words.
- Glossary from the text.

View chemical concepts

Explore molecular activity

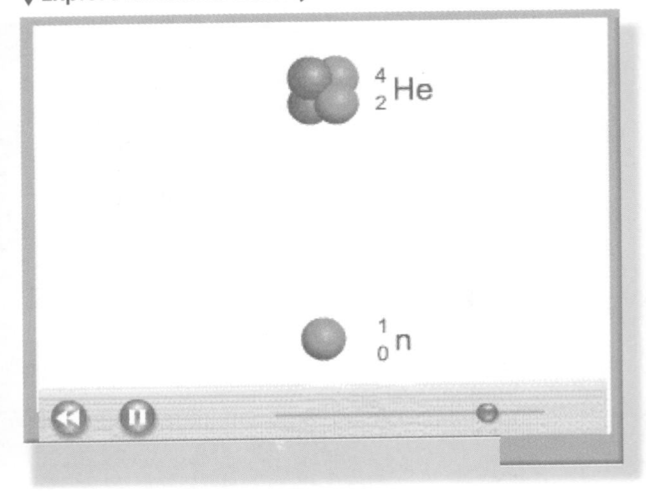

Visit: **chemistry.college.hmco.com/students**
For Technical Support: **(800) 732-3223** or **support@hmco.com**

Student CD-ROM

The following page explains how to launch and run the Student CD-ROM. After launching the program, please consult the Help section from the menu if you need more information on navigation or functionality. **Please see page 15 of this guide for the minimum system requirements.**

Installing the Student CD-ROM for Windows®

1. Insert the **Student CD-ROM** into your disk drive and close the door.
2. Double-click the **My Computer** icon on your desktop, then double-click the CD drive in which you placed the **Student CD-ROM**, and double-click the file "start.html".

From the Main Menu, select a chapter to begin. For more information on how to use the Student CD-ROM, see the ReadMe file located in the same place as the "start.html" file, or click Help, available from within the Student CD-ROM.

Note: QuickTime 6 for Windows and Shockwave Player must be installed on your computer.

To install QuickTime 6 for Windows and Shockwave Player:

1. Locate the QuickTime installer – QuickTimeInstaller.exe – on the Student CD-ROM in the folder labeled QuickTime Installer. Double-click this file and follow the on-screen instructions.
2. Locate the Shockwave installer – Shockwave_851_Installer.exe – on the Student CD-ROM in the folder labeled Shockwave Installer. Double-click this file and follow the on-screen instructions.

Note: If you cannot view the videos or animations in the Visualizations section, and you have installed QuickTime and Shockwave from this disk, you may have a problem with the number of versions of Shockwave installed on your computer.

1. From the Student CD-ROM, go to the QTXtra folder and copy the file QTXtra.32.
2. Double-click the My Computer icon on your desktop, then browse through to the directory C:\Windows\System\Macromed\Shockwave 8\Xtras and paste the file. If there is more than one Shockwave folder within the Macromed folder (Shockwave, Shockwave 8, etc.) paste the file in each Xtras folder. (Note: for Windows 2000, NT, and XP, the path to the Xtras folder is C:\WINNT\System32\Macromed\Shockwave 8\Xtras.).

Installing the Student CD-ROM for Macintosh®

1. Insert the **Student CD-ROM** into your disk drive and close the door.
2. Double-click the CD-ROM icon labeled **General Chemistry 8e** on your desktop
3. Double-click the file "start.html" if you are using Internet Explorer, or the file "startn.html" if you are using Netscape Navigator.

From the Main Menu, select a chapter to begin. For more information on how to use the Student CD-ROM, see the ReadMe file located in the same place as the "start.html" file, or click Help, available from within the Student CD-ROM.

Note: QuickTime 6 for Windows and Shockwave Player must be installed on your computer.

To install QuickTime 6 for Windows and Shockwave Player:

1. Locate the QuickTime Installer folder on the Student CD-ROM, and then open the folder of the operating system you are using (9.x or OS X). Double-click the file QuickTime Installer and follow the on-screen instructions.
2. Locate the Shockwave Installer folder on the Student CD-ROM, and then open the folder of the operating sytem you are using (9.x or OS X). Double-click the file Shockwave Installer and follow the on-screen instructions.

Note: If you cannot view the videos or animations in the Visualizations section, and you have installed QuickTime and Shockwave from this disk, you may have a problem with the number of versions of Shockwave installed on your computer.

1. From the Student CD-ROM, go to the QTXtra folder and copy the file QuickTime Asset.
2. Navigate through your hard drive to the directory System Folder:Extensions:Macromedia:Shockwave:Xtras and paste the file. If there is more than one Shockwave folder within the Macromedia folder, paste the file in each Shockwave: Xtras folder.

If you experience any problems launching or running the Student CD-ROM, refer to the ReadMe file for hints on improving the performance of your computer.

Program Help

Help is available by clicking on the Help button at the bottom of the screen. See also the file How to Use the CD on the root directory of the CD.

If the product is defective, contact Houghton Mifflin within 30 days of purchase at the number below. You will receive shipping instructions for the return and replacement of the defective disc(s). Provided the disc(s) has not been physically damaged, Houghton Mifflin will replace the disc(s).

Eduspace® (powered by Blackboard™)

Best Uses

For Students

- Complete assignments specifically chosen by your instructor
- Practice problem solving with online homework
- Access live online tutoring through *SMARTHINKING*™
- Visualize chemical concepts at the molecular level
- Review chapter concepts
- Access course materials and study strategies

Eduspace®(powered by Blackboard™), Houghton Mifflin's comprehensive course management solution, features an automatically graded online homework system with three types of problems: practice exercises based on in-text examples with rejoinders, selected end-of-chapter questions, and test bank questions.

Online homework

Blackboard Learning System TM (Release 6) - Microsoft Internet Explorer

File Edit View Favorites Tools Help

EDUspace®

Home Help Logout

HOUGHTON MIFFLIN

My Eduspace | Courses | General Resources | News | Support

Course Home
Course Policies
Course Materials
Tutoring
Communication
Course Tools

Course Map
Control Panel

Question 1 Multiple Choice — 1 points

What is the density of a solid object if 52.9 cm^3 of the solid has a mass of 16.2 g?

- ❍ 0.306 g/cm^3
- ❍ 0.341 g/cm^3
- ❍ 857 g/cm^3
- ❍ 1.17×10^3 g/cm^3
- ❍ 3.27 g/cm^3

Question 2 Multiple Choice — 1 points

When 21.0 grams of sodium metal is reacted with chlorine gas, 53.38 grams of table salt, sodium chloride, is formed. What is the mass of chlorine gas that reacted?

- ❍ 29.8 grams
- ❍ 29.3 grams
- ❍ 34.1 grams
- ❍ 32.4 grams
- ❍ 74.4 grams

Visit: **chemistry.college.hmco.com/students**

For Eduspace® Technical Support: **toll free (866) 817-3139**

Logging On

If your instructor has chosen to use *Eduspace®* to assign homework, tutorials, and other assignments via the Web, you can access your course by following the instructions in your *Eduspace®* Getting Started Guide for Students. You will need both the registration code in that guide and the course code given to you by your instructor in order to register.

For your reference

Periodic Table of the Elements

Alkali metals (group 1) · Alkaline earth metals (group 2) · Transition metals (groups 3–12) · Halogens (group 17) · Noble gases (group 18)

1 (1A)	2 (2A)	3	4	5	6	7	8	9	10	11	12	13 (3A)	14 (4A)	15 (5A)	16 (6A)	17 (7A)	18 (8A)
1 **H** 1.008																	2 **He** 4.003
3 **Li** 6.941	4 **Be** 9.012											5 **B** 10.81	6 **C** 12.01	7 **N** 14.01	8 **O** 16.00	9 **F** 19.00	10 **Ne** 20.18
11 **Na** 22.99	12 **Mg** 24.31											13 **Al** 26.98	14 **Si** 28.09	15 **P** 30.97	16 **S** 32.07	17 **Cl** 35.45	18 **Ar** 39.95
19 **K** 39.10	20 **Ca** 40.08	21 **Sc** 44.96	22 **Ti** 47.88	23 **V** 50.94	24 **Cr** 52.00	25 **Mn** 54.94	26 **Fe** 55.85	27 **Co** 58.93	28 **Ni** 58.69	29 **Cu** 63.55	30 **Zn** 65.38	31 **Ga** 69.72	32 **Ge** 72.59	33 **As** 74.92	34 **Se** 78.96	35 **Br** 79.90	36 **Kr** 83.80
37 **Rb** 85.47	38 **Sr** 87.62	39 **Y** 88.91	40 **Zr** 91.22	41 **Nb** 92.91	42 **Mo** 95.94	43 **Tc** (98)	44 **Ru** 101.1	45 **Rh** 102.9	46 **Pd** 106.4	47 **Ag** 107.9	48 **Cd** 112.4	49 **In** 114.8	50 **Sn** 118.7	51 **Sb** 121.8	52 **Te** 127.6	53 **I** 126.9	54 **Xe** 131.3
55 **Cs** 132.9	56 **Ba** 137.3	57 **La*** 138.9	72 **Hf** 178.5	73 **Ta** 180.9	74 **W** 183.9	75 **Re** 186.2	76 **Os** 190.2	77 **Ir** 192.2	78 **Pt** 195.1	79 **Au** 197.0	80 **Hg** 200.6	81 **Tl** 204.4	82 **Pb** 207.2	83 **Bi** 209.0	84 **Po** (209)	85 **At** (210)	86 **Rn** (222)
87 **Fr** (223)	88 **Ra** 226	89 **Ac†** (227)	104 **Rf** (261)	105 **Db** (262)	106 **Sg** (263)	107 **Bh** (264)	108 **Hs** (265)	109 **Mt** (268)	110 **Ds** (281)	111 **Uuu**	112 **Uub**		114 **Uuq**				

metals ← → nonmetals

*Lanthanides	58 **Ce** 140.1	59 **Pr** 140.9	60 **Nd** 144.2	61 **Pm** (145)	62 **Sm** 150.4	63 **Eu** 152.0	64 **Gd** 157.3	65 **Tb** 158.9	66 **Dy** 162.5	67 **Ho** 164.9	68 **Er** 167.3	69 **Tm** 168.9	70 **Yb** 173.0	71 **Lu** 175.0
†Actinides	90 **Th** 232.0	91 **Pa** (231)	92 **U** 238.0	93 **Np** (237)	94 **Pu** (244)	95 **Am** (243)	96 **Cm** (247)	97 **Bk** (247)	98 **Cf** (251)	99 **Es** (252)	100 **Fm** (257)	101 **Md** (258)	102 **No** (259)	103 **Lr** (260)

Group numbers 1–18 represent the system recommended by the International Union of Pure and Applied Chemistry.

For your reference

Table of Atomic Masses*

Element	Symbol	Atomic Number	Atomic Mass	Element	Symbol	Atomic Number	Atomic Mass	Element	Symbol	Atomic Number	Atomic Mass
Actinium	Ac	89	(227)†	Gold	Au	79	197.0	Praseodymium	Pr	59	140.9
Aluminum	Al	13	26.98	Hafnium	Hf	72	178.5	Promethium	Pm	61	(145)
Americium	Am	95	(243)	Hassium	Hs	108	(265)	Protactinium	Pa	91	(231)
Antimony	Sb	51	121.8	Helium	He	2	4.003	Radium	Ra	88	226
Argon	Ar	18	39.95	Holmium	Ho	67	164.9	Radon	Rn	86	(222)
Arsenic	As	33	74.92	Hydrogen	H	1	1.008	Rhenium	Re	75	186.2
Astatine	At	85	(210)	Indium	In	49	114.8	Rhodium	Rh	45	102.9
Barium	Ba	56	137.3	Iodine	I	53	126.9	Rubidium	Rb	37	85.47
Berkelium	Bk	97	(247)	Iridium	Ir	77	192.2	Ruthenium	Ru	44	101.1
Beryllium	Be	4	9.012	Iron	Fe	26	55.85	Rutherfordium	Rf	104	(261)
Bismuth	Bi	83	209.0	Krypton	Kr	36	83.80	Samarium	Sm	62	150.4
Bohrium	Bh	107	(264)	Lanthanum	La	57	138.9	Scandium	Sc	21	44.96
Boron	B	5	10.81	Lawrencium	Lr	103	(260)	Seaborgium	Sg	106	(263)
Bromine	Br	35	79.90	Lead	Pb	82	207.2	Selenium	Se	34	78.96
Cadmium	Cd	48	112.4	Lithium	Li	3	6.941	Silicon	Si	14	28.09
Calcium	Ca	20	40.08	Lutetium	Lu	71	175.0	Silver	Ag	47	107.9
Californium	Cf	98	(251)	Magnesium	Mg	12	24.31	Sodium	Na	11	22.99
Carbon	C	6	12.01	Manganese	Mn	25	54.94	Strontium	Sr	38	87.62
Cerium	Ce	58	140.1	Meitnerium	Mt	109	(268)	Sulfur	S	16	32.07
Cesium	Cs	55	132.9	Mendelevium	Md	101	(258)	Tantalum	Ta	73	180.9
Chlorine	Cl	17	35.45	Mercury	Hg	80	200.6	Technetium	Tc	43	(98)
Chromium	Cr	24	52.00	Molybdenum	Mo	42	95.94	Tellurium	Te	52	127.6
Cobalt	Co	27	58.93	Neodymium	Nd	60	144.2	Terbium	Tb	65	158.9
Copper	Cu	29	63.55	Neon	Ne	10	20.18	Thallium	Tl	81	204.4
Curium	Cm	96	(247)	Neptunium	Np	93	(237)	Thorium	Th	90	232.0
Darmstadtium	Ds	110	(281)	Nickel	Ni	28	58.69	Thulium	Tm	69	168.9
Dubnium	Db	105	(262)	Niobium	Nb	41	92.91	Tin	Sn	50	118.7
Dysprosium	Dy	66	162.5	Nitrogen	N	7	14.01	Titanium	Ti	22	47.88
Einsteinium	Es	99	(252)	Nobelium	No	102	(259)	Tungsten	W	74	183.9
Erbium	Er	68	167.3	Osmium	Os	76	190.2	Uranium	U	92	238.0
Europium	Eu	63	152.0	Oxygen	O	8	16.00	Vanadium	V	23	50.94
Fermium	Fm	100	(257)	Palladium	Pd	46	106.4	Xenon	Xe	54	131.3
Fluorine	F	9	19.00	Phosphorus	P	15	30.97	Ytterbium	Yb	70	173.0
Francium	Fr	87	(223)	Platinum	Pt	78	195.1	Yttrium	Y	39	88.91
Gadolinium	Gd	64	157.3	Plutonium	Pu	94	(244)	Zinc	Zn	30	65.38
Gallium	Ga	31	69.72	Polonium	Po	84	(209)	Zirconium	Zr	40	91.22
Germanium	Ge	32	72.59	Potassium	K	19	(39.10				

*The values given here are to four significant figures where possible.

†A value given in parentheses denotes the mass of the longest-lived isotope.

Print Supplements Available Separately

Student Solutions Manual (ISBN 0-618-39945-3)

David Bookin, *Mount San Jacinto College*, Darrell D. Ebbing, *Wayne State University*, and Steven D. Gammon, *Western Washington University*.

This manual contains detailed solutions to the in-chapter exercises and odd-numbered end-of-chapter problems. It also contains answers to the Review Questions. All solutions have been checked by an independent reviewer to ensure accuracy.

Study Guide for *General Chemistry* (ISBN 0-618-39943-7)

Larry K. Krannich, *University of Alabama at Birmingham*.

Each chapter of the study guide reinforces your understanding of concepts and operational skills presented in the text. It includes the following features for each chapter: a list of key terms and their definitions, a diagnostic test with answers, a summary of major concepts and operational skills, additional practice problems and their solutions, and a chapter post-test with answers.

To purchase these supplements, ask at your bookstore, visit our website at college.hmco.com, or call Houghton Mifflin Customer Service at **(800) 225-1464.**

Minimum System Requirements

Student CD-ROM

Windows®

- Windows® XP, 2000, 98, ME, NT 4.0 Service Pack 6
- Pentium II or higher
- 128 MB RAM, 256 MB RAM recommended
- 20 MB hard disk space
- Internet Explorer 6.x, 5.5, 5.0; Netscape 7.0, 6.2, 6.1
- Screen Display: 800 x 600 minimum, 1024 x 768 recommended
- Color Display: Thousands of colors
- CD-ROM drive
- Soundcard and Speakers
- Printer
- QuickTime® 6 (included)
- Shockwave 8.5 Player (included)

Macintosh®

- OSX 10.2, 10.1, 10.0, OS 9.2, 9.1
- PowerMac 68030 or higher
- 128 MB RAM minimum, 256 MB RAM recommended
- 20 MB hard disk space
- Internet Explorer 5.0; Netscape 7.0, 6.2
- Screen Display: 800 x 600 minimum, 1024 x 768 recommended
- Color Display: Thousands of colors
- CD-ROM drive
- Soundcard and Speakers
- Printer
- QuickTime® 6 (included)
- Shockwave 8.5 Player (included)

Eduspace® System Requirements

Minimum Requirements

- Microsoft Windows 2000 or Windows XP
  or
  Macintosh OS 9.2, 10.1 or 10.2
- Microsoft Internet Explorer 5.5 (or above)
  or
  Netscape Navigator 6.1 (or higher, 7.1 or higher for Windows XP or Macintosh OS 10.2)
- Acrobat Reader 4.0
- Display with 256 colors and 800x600 resolution or higher
- 56 K modem connection

Strongly Recommended Hardware/Software

- Microsoft 2000 with Microsoft Internet Explorer 6.0 or Netscape Navigator 6.1
  or
  Microsoft Windows XP with Internet Explorer 6.0 or Netscape Navigator 7.1
  or
  Macintosh OS 10.2 with Internet Explorer 5.2 or Netscape Navigator 7.0 (or higher)
- Adobe Acrobat 6.0
- Display with thousands of colors and 1024x768 resolution or higher
- T1 line, cable modem, DSL, or company Internet LAN

Note that Houghton Mifflin recommends using the Sun Java Run-time Environment (JRE) 1.4.x in your Web browser. Use of the Microsoft JVM is not recommended.

Visit: **chemistry.college.hmco.com/students**

For Technical Support: **(800) 732-3223** or **support@hmco.com**

Ever imagine yourself working as a high school teacher?

Or maybe as a chemist for the Federal Bureau of Investigation?

Or perhaps as a patent attorney for a corporation?

Put your chemistry knowledge to work!

Many of these jobs require a background in chemistry. Remember to check the *Student Website* for the latest information on careers in chemistry.